I0062040

THIS BOOK BELONGS TO

DEDICATION

This Farm Record Keeping Log Book is dedicated to farmers and ranchers who want to manage and track livestock, record equipment maintenance and keep accurate expense and income records.

You are my inspiration for producing this book and I'm honored to be a part of helping you manage and retain important information regarding your farming business.

HOW TO USE THIS BOOK

This Farm Record Book will help you record, collect, and organize your information in an easy to use format.

Here are examples of information for you to fill in and write the details of your farming business.

Fill in the following information:

1. Business Details - name, address, email address, phone, fax, website, logbook (start and end dates), and notes
2. Livestock Record - date, type of livestock, start date, end date, weight average, value, and balance
3. Equipment Maintenance and Repair - date, equipment, inspections, repairs, service repair dates, remarks
4. Notes - full page each month for notes
5. Farm Expenses- 12-month expense pages: date, expenses, cost remarks
6. Income - 12-month income pages: date, source, description, payment method, amount

BUSINESS DETAILS

BUSINESS NAME:	
ADDRESS:	

EMAIL ADDRESS:	PHONE NUMBER:
FAX NUMBER:	WEBSITE:

LOGBOOK DETAILS

CONTINUED FROM LOGBOOK:	CONTINUED TO LOG BOOK:
DATE LOG STARTED:	DATE LOG ENDED:

NOTES

LIVESTOCK RECORD

DATE:		START DATE:			SEND OF YEAR:			
NO	TYPE OF LIVESTOCK	QTY	AVG WEIGHT	VALUE	QTY	AVG WEIGHT	VALUE	BALANCE

EQUIPMENT MAINTENANCE & REPAIR

MONTH:	

DATE	EQUIPMENT	INSPECTION / MAINTENANCE / REPAIR / SERVICES REQUIRED	SERVICE / REPAIR DATE	INITIALS	REMARKS

NOTES

FARM EXPENSES

MONTH:			
DATE	EXPENSES	COST	REMARKS

FARM EXPENSES

MONTH:			
DATE	EXPENSES	COST	REMARKS

FARM EXPENSES

MONTH:			
DATE	EXPENSES	COST	REMARKS

FARM INCOME

MONTH:				
DATE	SOURCE	DESCRIPTION	PAYMENT METHOD	AMOUNT

FARM INCOME

MONTH:				
DATE	SOURCE	DESCRIPTION	PAYMENT METHOD	AMOUNT

LIVESTOCK RECORD

DATE:		START DATE:			SEND OF YEAR:			
NO	TYPE OF LIVESTOCK	QTY	AVG WEIGHT	VALUE	QTY	AVG WEIGHT	VALUE	BALANCE

EQUIPMENT MAINTENANCE & REPAIR

MONTH:					
DATE	EQUIPMENT	INSPECTION / MAINTENANCE / REPAIR / SERVICES REQUIRED	SERVICE / REPAIR DATE	INITIALS	REMARKS

NOTES

FARM EXPENSES

MONTH:			
DATE	EXPENSES	COST	REMARKS

FARM EXPENSES

MONTH:			
DATE	EXPENSES	COST	REMARKS

FARM EXPENSES

MONTH:

DATE	EXPENSES	COST	REMARKS

FARM INCOME

MONTH:

DATE	SOURCE	DESCRIPTION	PAYMENT METHOD	AMOUNT

FARM INCOME

MONTH:				
DATE	SOURCE	DESCRIPTION	PAYMENT METHOD	AMOUNT

LIVESTOCK RECORD

DATE:		START DATE:			SEND OF YEAR:			
NO	TYPE OF LIVESTOCK	QTY	AVG WEIGHT	VALUE	QTY	AVG WEIGHT	VALUE	BALANCE

EQUIPMENT MAINTENANCE & REPAIR

MONTH:					
DATE	EQUIPMENT	INSPECTION / MAINTENANCE / REPAIR / SERVICES REQUIRED	SERVICE / REPAIR DATE	INITIALS	REMARKS

NOTES

FARM EXPENSES

MONTH:			
DATE	EXPENSES	COST	REMARKS

FARM EXPENSES

MONTH:			
DATE	**EXPENSES**	**COST**	**REMARKS**

FARM EXPENSES

MONTH:			
DATE	**EXPENSES**	**COST**	**REMARKS**

FARM INCOME

MONTH:				
DATE	SOURCE	DESCRIPTION	PAYMENT METHOD	AMOUNT

FARM INCOME

MONTH:				
DATE	SOURCE	DESCRIPTION	PAYMENT METHOD	AMOUNT

LIVESTOCK RECORD

DATE:		START DATE:			SEND OF YEAR:			
NO	TYPE OF LIVESTOCK	QTY	AVG WEIGHT	VALUE	QTY	AVG WEIGHT	VALUE	BALANCE

EQUIPMENT MAINTENANCE & REPAIR

MONTH:					
DATE	EQUIPMENT	INSPECTION / MAINTENANCE / REPAIR / SERVICES REQUIRED	SERVICE / REPAIR DATE	INITIALS	REMARKS

NOTES

FARM EXPENSES

MONTH:

DATE	EXPENSES	COST	REMARKS

FARM EXPENSES

MONTH:			
DATE	**EXPENSES**	**COST**	**REMARKS**

FARM EXPENSES

MONTH:			
DATE	EXPENSES	COST	REMARKS

FARM INCOME

MONTH:				
DATE	SOURCE	DESCRIPTION	PAYMENT METHOD	AMOUNT

FARM INCOME

MONTH:				
DATE	SOURCE	DESCRIPTION	PAYMENT METHOD	AMOUNT

LIVESTOCK RECORD

DATE:		START DATE:			SEND OF YEAR:			
NO	TYPE OF LIVESTOCK	QTY	AVG WEIGHT	VALUE	QTY	AVG WEIGHT	VALUE	BALANCE

EQUIPMENT MAINTENANCE & REPAIR

MONTH:

DATE	EQUIPMENT	INSPECTION / MAINTENANCE / REPAIR / SERVICES REQUIRED	SERVICE / REPAIR DATE	INITIALS	REMARKS

NOTES

FARM EXPENSES

MONTH:			
DATE	EXPENSES	COST	REMARKS

FARM EXPENSES

MONTH:			
DATE	EXPENSES	COST	REMARKS

FARM EXPENSES

MONTH:			
DATE	EXPENSES	COST	REMARKS

FARM INCOME

MONTH:				
DATE	SOURCE	DESCRIPTION	PAYMENT METHOD	AMOUNT

FARM INCOME

MONTH:				
DATE	SOURCE	DESCRIPTION	PAYMENT METHOD	AMOUNT

LIVESTOCK RECORD

DATE:		START DATE:			SEND OF YEAR:			
NO	TYPE OF LIVESTOCK	QTY	AVG WEIGHT	VALUE	QTY	AVG WEIGHT	VALUE	BALANCE

EQUIPMENT MAINTENANCE & REPAIR

MONTH:

DATE	EQUIPMENT	INSPECTION / MAINTENANCE / REPAIR / SERVICES REQUIRED	SERVICE / REPAIR DATE	INITIALS	REMARKS

NOTES

FARM EXPENSES

MONTH:			
DATE	EXPENSES	COST	REMARKS

FARM EXPENSES

MONTH:			
DATE	EXPENSES	COST	REMARKS

FARM EXPENSES

MONTH:

DATE	EXPENSES	COST	REMARKS

FARM INCOME

MONTH:				
DATE	SOURCE	DESCRIPTION	PAYMENT METHOD	AMOUNT

FARM INCOME

MONTH:

DATE	SOURCE	DESCRIPTION	PAYMENT METHOD	AMOUNT

LIVESTOCK RECORD

DATE:		START DATE:			SEND OF YEAR:			
NO	TYPE OF LIVESTOCK	QTY	AVG WEIGHT	VALUE	QTY	AVG WEIGHT	VALUE	BALANCE

EQUIPMENT MAINTENANCE & REPAIR

MONTH:					
DATE	EQUIPMENT	INSPECTION / MAINTENANCE / REPAIR / SERVICES REQUIRED	SERVICE / REPAIR DATE	INITIALS	REMARKS

NOTES

FARM EXPENSES

MONTH:			
DATE	EXPENSES	COST	REMARKS

FARM EXPENSES

MONTH:			
DATE	EXPENSES	COST	REMARKS

FARM EXPENSES

MONTH:

DATE	EXPENSES	COST	REMARKS

FARM INCOME

MONTH:				
DATE	SOURCE	DESCRIPTION	PAYMENT METHOD	AMOUNT

FARM INCOME

MONTH:

DATE	SOURCE	DESCRIPTION	PAYMENT METHOD	AMOUNT

LIVESTOCK RECORD

DATE:		START DATE:			SEND OF YEAR:			
NO	TYPE OF LIVESTOCK	QTY	AVG WEIGHT	VALUE	QTY	AVG WEIGHT	VALUE	BALANCE

EQUIPMENT MAINTENANCE & REPAIR

MONTH:

DATE	EQUIPMENT	INSPECTION / MAINTENANCE / REPAIR / SERVICES REQUIRED	SERVICE / REPAIR DATE	INITIALS	REMARKS

NOTES

FARM EXPENSES

MONTH:			
DATE	EXPENSES	COST	REMARKS

FARM EXPENSES

MONTH:			
DATE	EXPENSES	COST	REMARKS

FARM EXPENSES

MONTH:			
DATE	EXPENSES	COST	REMARKS

FARM INCOME

MONTH:

DATE	SOURCE	DESCRIPTION	PAYMENT METHOD	AMOUNT

FARM INCOME

MONTH:

DATE	SOURCE	DESCRIPTION	PAYMENT METHOD	AMOUNT

LIVESTOCK RECORD

DATE:		START DATE:			SEND OF YEAR:			
NO	TYPE OF LIVESTOCK	QTY	AVG WEIGHT	VALUE	QTY	AVG WEIGHT	VALUE	BALANCE

EQUIPMENT MAINTENANCE & REPAIR

MONTH:					
DATE	EQUIPMENT	INSPECTION / MAINTENANCE / REPAIR / SERVICES REQUIRED	SERVICE / REPAIR DATE	INITIALS	REMARKS

NOTES

FARM EXPENSES

MONTH:			
DATE	EXPENSES	COST	REMARKS

FARM EXPENSES

MONTH:

DATE	EXPENSES	COST	REMARKS

FARM EXPENSES

MONTH:

DATE	EXPENSES	COST	REMARKS

FARM INCOME

MONTH:

DATE	SOURCE	DESCRIPTION	PAYMENT METHOD	AMOUNT

FARM INCOME

MONTH:				
DATE	**SOURCE**	**DESCRIPTION**	**PAYMENT METHOD**	**AMOUNT**

LIVESTOCK RECORD

DATE:		START DATE:			SEND OF YEAR:			
NO	TYPE OF LIVESTOCK	QTY	AVG WEIGHT	VALUE	QTY	AVG WEIGHT	VALUE	BALANCE

EQUIPMENT MAINTENANCE & REPAIR

MONTH:					
DATE	EQUIPMENT	INSPECTION / MAINTENANCE / REPAIR / SERVICES REQUIRED	SERVICE / REPAIR DATE	INITIALS	REMARKS

NOTES

FARM EXPENSES

MONTH:			
DATE	EXPENSES	COST	REMARKS

FARM EXPENSES

MONTH:			
DATE	EXPENSES	COST	REMARKS

FARM EXPENSES

MONTH:			
DATE	EXPENSES	COST	REMARKS

FARM INCOME

MONTH:				
DATE	SOURCE	DESCRIPTION	PAYMENT METHOD	AMOUNT

FARM INCOME

MONTH:				
DATE	SOURCE	DESCRIPTION	PAYMENT METHOD	AMOUNT

LIVESTOCK RECORD

DATE:		START DATE:			SEND OF YEAR:			
NO	TYPE OF LIVESTOCK	QTY	AVG WEIGHT	VALUE	QTY	AVG WEIGHT	VALUE	BALANCE

EQUIPMENT MAINTENANCE & REPAIR

MONTH:					
DATE	EQUIPMENT	INSPECTION / MAINTENANCE / REPAIR / SERVICES REQUIRED	SERVICE / REPAIR DATE	INITIALS	REMARKS

NOTES

FARM EXPENSES

MONTH:			
DATE	**EXPENSES**	**COST**	**REMARKS**

FARM EXPENSES

MONTH:			
DATE	EXPENSES	COST	REMARKS

FARM EXPENSES

MONTH:

DATE	EXPENSES	COST	REMARKS

FARM INCOME

MONTH:				
DATE	SOURCE	DESCRIPTION	PAYMENT METHOD	AMOUNT

FARM INCOME

MONTH:				
DATE	SOURCE	DESCRIPTION	PAYMENT METHOD	AMOUNT

LIVESTOCK RECORD

DATE:		START DATE:			SEND OF YEAR:			
NO	TYPE OF LIVESTOCK	QTY	AVG WEIGHT	VALUE	QTY	AVG WEIGHT	VALUE	BALANCE

EQUIPMENT MAINTENANCE & REPAIR

MONTH:					
DATE	EQUIPMENT	INSPECTION / MAINTENANCE / REPAIR / SERVICES REQUIRED	SERVICE / REPAIR DATE	INITIALS	REMARKS

NOTES

FARM EXPENSES

MONTH:			
DATE	EXPENSES	COST	REMARKS

FARM EXPENSES

MONTH:			
DATE	EXPENSES	COST	REMARKS

FARM EXPENSES

MONTH:			
DATE	EXPENSES	COST	REMARKS

FARM INCOME

MONTH:				
DATE	SOURCE	DESCRIPTION	PAYMENT METHOD	AMOUNT

FARM INCOME

MONTH:				
DATE	SOURCE	DESCRIPTION	PAYMENT METHOD	AMOUNT

www.ingramcontent.com/pod-product-compliance
Lightning Source LLC
Chambersburg PA
CBHW051759200326
41597CB00025B/4617
9781953557179